Lost, Stolen or Strayed?
Gordon Brown as Socialist Pretender
Ken Coates
Michael Meacher
MP for Oldham West and Royton

In what has been widely interpreted as a launch to his campaign to be elected as Tony Blair's successor in the Labour Party leadership, Gordon Brown went to the Mansion House on the 21st June 2006, to tell the City that he was a dependable upholder of 'stability and security'. His speech attracted the headlines because he threw his weight behind Tony Blair's plans to replace the submarine-based Trident nuclear weapon. We shall publish a considered view of this project shortly, but it is sufficient here to say that it is less of a military commitment than a declaration of undying subservience to the United States of America, which will exercise the most powerful control over its own weapons, when it subcontracts them to the British.

For the young Scots who are demonstrating against Trident, and mobilising for independent policies, it will be surprising that Gordon Brown was thought to have a reputation which was at times subversive, not only in favour of nuclear disarmament, but also ardently supporting the underdog and trade unions in Britain.

However, there was once a different Gordon, who will be able, when he meets his maker, to plead, in extenuation, previous good behaviour. Thus back in 1984 he told the House of Commons that:

> 'The dominant theme of this debate has been the concern expressed by Hon. Members about the escalating cost of the Trident programme, a project which is unacceptably expensive, economically wasteful and militarily unsound. It is a project which, while escalating the risks of nuclear war, puts at risk the integrity of our conventional defences. It is a project the implications of which in cost and security are not fully disclosed in the defence Estimates.'

Ten years earlier, before the rebirth of the movement for nuclear disarmament in the 'eighties with the campaign for European Nuclear Disarmament, Gordon drew together a comprehensive critique of policy and programme for change in Scotland, *The Red Paper on Scotland – The Socialist Challenge*. This was at one time required reading among the Scottish Left. Gordon Brown might usefully be reminded of his youthful thoughts which were generated at a time when he was innocent of experience on the Government's front bench, but strongly influenced by Scottish trade unions.

Unfortunately, the position has changed considerably. At present it is graphically summed up by Michael Meacher. As he explains (in *The Guardian*, 22nd June 2006) the condition of trade unionism hardly matches the expectations of *The Red Paper*.

> 'Gordon Brown last night again extolled the virtues of globalisation and Britain's achievements within it. In his Mansion House speech the Chancellor trumpeted "the most global and outward-looking of nations", while condemning any "sheltering against global competition". The benefits of steady and continuing growth are undeniable, but that is only part of the story.
>
> Britain has some of the worst working conditions in Europe. Britons work the longest hours in Europe. The Prime Minister continues to insist on an opt-out from the EU directive that limits the normal working week to 48 hours, and an exemption from the EU directive providing the same protection for agency workers as for their staff counterparts – a provision of particular importance for women. British workers cannot make a case for unfair dismissal until they have worked for an employer for a year – if sacked after 11 months they have no right of redress. If workers take legal industrial action, they have no protection – whatever the rights of their case – from being collectively dismissed after eight weeks.
>
> Trade union recognition rights are still denied to the 6 million employees – nearly a quarter of the workforce – who work in companies with 20 or fewer employees. This exemption of small firms from most employment protection legislation is not found elsewhere. Yet it is in these small firms, which constitute 85% of Britain's employers, that protection is most needed. These workplaces often have the worst health and safety records. And they employ a greater proportion of women and black people, on lower pay and subject to more discrimination. This exclusion really matters when Britain is the only country in the EU with no proper inspectorate of working conditions, yet prosecutions of employers are rare and there is no other mechanism for the exploited to seek protection.
>
> But it is part of a strategy to appease business and make the country attractive to foreign investment. In 1997 Tony Blair promised that "the changes that we propose would leave British law the most restrictive on trade unions in the western world". The Government's own trade and investment website takes the boast further: "UK law does not oblige employers to provide a written employment contract", and "Recruitment

costs in the UK are low ... and the law governing the conduct of employment agencies is less restrictive in the UK."

Easy hiring and firing is seen by the Government as a major selling point for companies, whatever the costs for workers' insecurity and powerlessness.

But it is a counter-productive policy. Even leaving aside the obvious injustices of the current approach and Britain's defiance of the International Labour Organisation convention, it has not improved productivity or competitiveness. Recent figures show that British productivity is far behind that of our closest competitors – 13% lower than Germany and 21% behind France. Moreover, productivity in the UK has been raised almost wholly by shedding labour – arguably the worst failing of the British economy over the past three decades. And we have slipped down the competitiveness league from fourth a decade ago to 11th now.

Britain's unemployment rate is half that of France and Germany. But that is largely due not to the absence of stifling social protection, as neo-liberals claim, but to the freeing up of monetary policy after Britain was ignominiously pitched out of the Exchange Rate Mechanism in 1992, and by staying outside the straitjacket of the one-size-fits-all Eurozone. Even so, the increasing Americanisation of the labour market over the past decade has not prevented a rapid decline in the private manufacturing sector with the loss of nearly a million jobs, while a huge unskilled and poorly educated workforce remains and unemployment is once again on the rise.

What is so tragic is that there is abundant evidence that greater protection at work, contrary to CBI received wisdom, benefits business and raises productivity, while instability in "flexible" markets does not. Precarious work, low pay, poor working conditions and long hours – Britain has the least regulated labour market, even including the US – undermine productivity, reduce motivation and increase absenteeism. That insecurity is aggravated by the lack of an international level playing field, which means British workers are cheaper to lay off for any multinational company looking to rationalise its workforce in Europe.

It isn't as though there is no proven alternative. The Nordic countries offer much better working conditions with no downside in economic performance – quite the reverse. Sweden matches Britain in growth, GDP per capita and unemployment level, and has a current account trade surplus of £5bn, against Britain's £40bn deficit. Even by New Labour's neo-liberal criteria, Sweden wins: it has lower inflation, higher global competitiveness and a better business record for creativity and research. And in quality of life, it's streets ahead. Its life expectancy is much higher, its poverty level is less than half that of Britain, its illiteracy rate is a third of ours, and its social mobility is far higher.

What, then, would workplace justice mean in Britain today? The national minimum wage should be a living wage, not (as at present) a poverty wage. The Warwick agreement on better working conditions, reached with the unions two years ago, should be implemented urgently, not put on a permanent backburner. A worker who has been found by a tribunal to have been unfairly dismissed should be entitled to reinstatement. Staff who want their union recognised should have to achieve a majority in a ballot, not be required (as at present) also to obtain support from at least 40% of everyone entitled to vote – no Government would accept a hurdle like that in getting itself elected.

When 400 workers are killed each year at work, health and safety should be much more rigorously enforced, with custodial penalties where gross managerial negligence is proven. The case for sympathetic action is clear and should be respected – Gate Gourmet workers would never have won justice if they had not been supported by workers in related jobs. And as is the law in many other European countries, employers should not be permitted to sack workers on a lawful strike. The Prime Minister promised "fairness not favours". We should demand nothing less.'

The Red Paper on Scotland
The Socialist Challenge
by Gordon Brown

The irresistible march of recent events places Scotland today at a turning point – not of our own choosing but where a choice must sooner or later be made. A resurgent nationalism which forces on to the agenda the most significant constitutional decisions since the Act of Union is one aspect of what even the *Financial Times* has described as 'a revolt of rising expectations.' But the proliferation of industrial unrest and the less publicised mushrooming of community action also bear witness to the sheer enormity of the gap now growing between people's conditions of living and their legitimate aspirations.

Yet the great debate on Scotland's future ushered in by the Kilbrandon report and precipitated by North Sea oil and Britain's economic crisis has hardly been a debate at all. Dominated by electoral calculations, nationalist and anti-nationalist passions and crude bribery, it has engendered a barren, myopic, almost suffocating consensus which has tended to ignore Scotland's real problems – our unstable economy and unacceptable level of unemployment, chronic inequalities of wealth and power and inadequate social services. And while Kilbrandon identified 'a diffuse feeling of dissatisfaction ... a feeling of powerlessness at the we/they relationship', the basic questions which face the Scotland of the nineteen-eighties remain unasked as well as unanswered: who shall exercise power and control the lives of our people? How can we harness our material resources and social energies to meet the needs of five million people and more? What social structure can guarantee to people the maximum control and self-management over the decisions which affect their lives, allowing the planned co-ordination of the use and distribution of resources, in a co-operative community of equals?

Scotland's social condition and political predicament cries out for a new commitment to socialist ideals, policies and action emerging from a far-reaching analysis of economy and society; a bringing together of the many positive insights, responses, and analyses to break through the deliberate separation of issues and the consequent fragmentation of people's consciousness; and a searching for a new social vision for Scotland which begins from people's potentials, is sensitive to cultural needs, and is humane, democratic and revolutionary. What the *Red*

Paper[1] – a contribution of many socialist views to such a debate – seeks to do is to transcend that false and sterile antithesis which has been manufactured between the nationalism of the SNP and the anti-nationalism of the Unionist parties, by concentrating on the fundamental realities of inequality and irresponsible social control, of private power and an inadequate democracy. For when the question of freedom for Scotland is raised, we must ask: freedom for whom? From what? For what?

Two themes are integral and complementary in the essays. The first is that the social and economic problems confronting Scotland arise not from national suppression nor from London mismanagement (although we have had our share of both) but from the uneven and uncontrolled development of capitalism, and the failure of successive governments to challenge and transform it. Thus we cannot hope to resolve such problems merely by recovering a lost independence or through inserting another tier of government: what is required is planned control of our economy and a transformation of democracy at all levels. The second theme is more basic than that. We suggest the real resources of Scotland are not the reserves of oil beneath the sea (nor the ingenuity of native entrepreneurs) but the collective energies and potential of our people whose abilities and capacities have been stultified by a social system which has for centuries sacrificed social aspirations to private ambitions. It is argued that what appear to be contradictory features of Scottish life today – militancy and apathy, cynicism and a thirst for change – can best be understood as working people's frustration with and refusal to accept powerlessness and lack of control over blind social forces which determine their lives. It is a disenchantment which underlines an untapped potential for co-operative action upon which we must build.

The vision of the early socialists was of a society which had abolished for ever the dichotomy – the split personality caused by people's unequal control over their social development – between man's personal and collective existence, by substituting communal co-operation for the divisive forces of competition. Today the logic of present economic development, in inflation and stagnation, and at the same time the demand for the fullest use of material resources, makes it increasingly impossible to manage the economy both for private profit and the needs of society as a whole. Yet the long-standing paradox of Scottish politics has been the surging forward of working class industrial and political pressure (and in particular the loyal support given to Labour) and its containment through the accumulative failures of successive Labour Governments. More than fifty years ago socialism was a qualitative concept, an urgently felt moral imperative, about social control (and not merely state control or more or less equality). Today for many it means little more than a scheme for compensating the least fortunate in an unequal society. We suggest that the rise of modern Scottish nationalism is less an assertion of Scotland's permanence as a nation than a response to Scotland's uneven development – in particular to the gap between people's experiences as part of an increasingly demoralised Great Britain and their (oil-fired) expectations at a Scottish level. Thus, the discontent is a measure of both Scottish and British socialists to advance far and fast enough in shifting the balance of wealth and power to working people and in raising people's awareness – especially outside the central belt of Scotland in areas where inequalities are greater – about the co-operative possibilities for modern society.

Clearly it is easy to overestimate the potential for radical social change in Scotland today – but it is dangerous to underestimate the demand for it. For the first time since the Union, oil and the political response to it has swung the balance of influence within Great Britain in favour of Scotland, giving the Scottish Labour Movement in particular a new bargaining power. Labour has two choices: to deflect the present discontent, resisting the pressure for change until it becomes inevitable, at the risk of the success of an SNP which presumes the familiar priorities of wealth and power set over people, or to harness the wide ranging dissatisfaction in a socialist strategy which not only forces the pace of the advance towards socialism in Britain as a whole, but seeks to revitalise the grass roots of Scottish society. This means drawing new links between political action and the various positive movements for control in community and industrial life, to forge a new kind of social and economic democracy. Such a strategy implies that Scottish socialists cannot support a strategy for independence which postpones the question of meeting urgent social and economic needs until the day after independence – but nor can they give unconditional support to maintaining the integrity of the United Kingdom – and all that that entails – without any guarantee of radical social change; the question is not one of structures nor of territorial influence, but of democracy – how working people in Scotland can increase the control they have over the decisions which shape their lives and the wealth they alone produce – and in doing so aid the struggle for a shift of power to working people elsewhere.

Social Needs

Any study of Scotland today must start from where people are, the realities of day-to-day living, extremes of wealth and poverty, unequal opportunities at work, in housing, health, education and community living generally. The gross inequalities which disfigure Scottish social life (and British society as a whole) have been obscured by a debate which merely poses the choice between separatism and unionism. For there are rich Scots, very rich Scots – and very poor Scots. The Sample Census of 1966 confirms that Scotland has a small owning and managing élite, a small but increasingly important professional and supervisory class, and a very large working class embracing over 80 per cent of Scottish people, with the balance shifting from skilled manual to service workers[2]. In drawing attention to the fact that average holdings of equities were twice as high in Scotland than England and that Scotland had 11 per cent of British people with assets above £40,000, *The Economist* recently suggested the extremes of wealth were more pronounced in Scotland than in Britain as a whole (and by implication in Western Europe and America)[3]. In 1971-72 there were 664 estates above £50,000 and 182 above £100,000 (with an average of two wills above £1m for each of the previous ten years)[4]. Ian Levitt draws attention to the marked inequalities in income which do not arise from differing skills, contributions to community life or needs. The top two per cent earn seven per cent of total income (more than half of that coming from unearned income in investments and profits), the top five per cent earn fifteen per cent of income and the top twelve per cent of Scottish people earn as much as the bottom fifty per cent. Another dimension of inequality is explored by John McEwen who provides new and important evidence of the concentration of land ownership in the Highlands and Islands where 340 individuals (or companies) own sixty-four per cent (over six million acres) of the land. Four individuals own one-sixteenth, ten own one-eighth, and forty own one-third and one hundred and fifty own one-half of the Highlands.

There are three distinguishing marks of the new structure of inequality in Scotland – the failure of taxation to erode the power of private property (those earning below £2,000 in 1971 paid half income tax)[5], the dramatic growth of private occupational and pensions schemes to create a new structure of privilege within our social security system, and the sheer extent of poverty itself. Levitt estimates that 23 per cent of Scottish people are living at or just below the poverty line set by the Supplementary Benefits Commission – one and a quarter million people concentrated in four large groups, old people, low paid workers, and their families, the unemployed and single parent families. Supplementary benefits payments have trebled since 1947 and doubled over the last decade. In showing how the poverty trap affects localities as it does individuals, Vincent Cable draws attention to the extent of multiple deprivation in areas of Glasgow. While Clydeside is the heart of the problem, the recent Bowhouse study which showed 90 per cent of families in poverty, 40 per cent of active males unemployed, and 90 per cent of child school-leavers without 'O' of 'H' levels in one area of Clackmannanshire, emphasises that community deprivation is nation-wide[6]. Richard Bryant demonstrates that it would take an extra expenditure of more than 41p per person yearly to raise Scottish social services merely to the average provision of England and Wales, but the root of the problem, which cannot be solved by better or more casework, or even community action linked to corporate planning, is, he suggests, 'the structural nature of poverty within an unequal society which rewards its members according to the principles of a market economy and not on the basis of human need.' For the inequalities which distort Scottish life are generated from the work-place outwards: low pay, placing at least 100,000 male workers and their families on the poverty line, insufficient provision for retirement and for families deprived of their breadwinner, and the threat and actuality of redundancy. With Scotland's unemployment figures remaining up to three time as high as the South East of England even in periods of boom, about one-third of our unemployed have become long term out-of-work and only 36 per cent of all unemployed receive insurance benefits.

Our other contributors show that although public expenditure in Scotland is 17 per cent higher per person than in Britain as a whole[7], existing social service provision tends to mirror rather than redress existing inequalities. While there have been national appeals to Scotland's inherent capacity for democracy and equality, and our passion for social justice, it cannot be said that the controls exercised in Scotland over local authorities and social institutions – education, law and religion – have led to a searching for different concepts of welfare or standards of need. Today's poor remain stigmatised by poverty not simply because social service provision is inadequate, but because there is no dominant concept of reciprocity in our social services. The failure of the Dundee Education Priority experiment is symptomatic of how little enthusiasm there has been from local administrators and politicians for projects providing greater resources to areas of need[8]. Robin Cook points to the inadequacies of existing housing policies. One in ten houses in Scotland are substandard, 'most of them if not all', concluded the Scottish Development Department, 'could have been closed as unfit fifty years ago' – and despite a massive inflow of public funds into the private sector (over 40 per cent of Scottish homes are now owner-occupied compared with 25 per cent in 1964), only 15 per cent of working class families are owner-occupiers (most of them living in houses built before 1919). Fewer are now able to buy their own home.[9] Donald Cameron draws attention to the prevalence of higher rates of infant mortality in Scotland – the most sensitive index of public health – with mortality among infants, as it was twenty years ago, twice as high among social classes 4 and 5 as it is among classes 1 and 2. While Richard Bryant and Colin Kirkwood draw attention to the paltry provision of social service and community facilities, Nigel Grant emphasises that the substantially greater resources devoted to education – ten times those allocated to social work – have not been the force for social equality presumed in our traditional assumptions about Scottish education or envisaged in the British reforms of the sixties. Only one in twelve of 3-5 year olds enjoy nursery provision (half the rate for Britain as a whole), 55 per cent of children leave school without 'O' or 'H' levels, 22 per cent of school leavers enjoy the day or block release rights legislated in 1944; and the proportion of the sons and daughters of working class parents among students in Scottish universities – again a most sensitive index of inequality – has been declining rather than increasing, probably to a level below that of the nineteen twenties[10]. Those who argue that Scotland's democratic traditions in education can be a powerful lever for redressing inequality and regenerating Scottish society are left with the sole alternative of supporting policies for mobilising social and economic as well as educational resources in an overall strategy for ending inequality.

Community Democracy

Scotland desperately needs a widely articulated and sufficiently popular concept of welfare and need grounded in inequality and reciprocity in framing social policies and social priorities. Bryant, Cameron, Grant, Cook and Levitt all point to what is urgently required merely to meet people's elemental needs – a massive expansion in housing and community amenities, a regeneration of the public sector, and improvement of public health facilities especially in the community and industrial health fields, greater concentration of educational resources among those with the least opportunities, and a phasing out of means-tested benefits by adequate provision by right for the old, the single parent family, the unemployed, the disabled, and the low paid. But from community action groups – tenants associations, organisations of the unemployed, the old, the homeless, and the sick, movements to fight oil-related developments, anti-social planning decisions and so on – to the activities of specialised pressure groups and professional social workers, teachers and health workers themselves, the demand is increasingly that society be organised in a manner to cater for people's needs, that community goals be set to meet people's requirements as they express them. If the prospects for the least fortunate are to be as great as they can be, then they must have the final say – and that requires a massive and irreversible shift of power to working people, a framework of free universal welfare services controlled by the people who use them. This potential for community action in harnessing local planning community initiatives is emphasised by Ronald Young in his study of the inadequacies of local government. But he also shows that the challenge to the Scottish Labour Movement is to stimulate and co-ordinate movements towards linking what are at present sectional movements so that they involve all the poor, all the badly housed and all the deprived, and towards making more explicit social priorities and social needs from the community outwards. A first step could be compiling on a nation-wide basis an inventory of social needs, 'a social audit'[11] prepared by community groups themselves. But socialism will have to be won also at the point of production – the production of needs, ideas and particularly of goods and services. And that demands ending the power of a minority through ownership and control to direct the energies of all other members of our society.

Economic Prospects

Three facts stand out from a study of the present Scottish economy: the failure to create much needed jobs and so to eliminate the disparities between Scottish and UK rates of unemployment; the inability to develop a new economic base for sustained economic growth; and the increased level of external control over the Scottish economy. That over a quarter of a million new jobs are still required over the next decade is a measure of the failure of existing economic strategies. Today the clear industrial divide is between the giants of finance and industry on an international scale (including British multinationals, second only to the Americans) and the small firms on the national market. In presenting new and important evidence, John Firn shows the extent to which the main productive sector of the Scottish economy, manufacturing industry, employing one third of Scottish workers, is both externally and monopolistically controlled.

Fifty nine per cent of manufacturing workers are employed by companies based outside Scotland and the larger the enterprise and the faster growing it is, the more likely is external control. Moreover, one hundred and ten enterprises account for nearly half of manufacturing employment, three quarters of firms with more than 5,000 employees being externally controlled and the top twenty British multinationals alone employing one in twelve of manufacturing workers in Scotland. Twice as much per capita is invested in Scotland by American capital than in Britain as a whole, Scotland being the largest area per head of population for American intervention after Canada. It is, Firn concludes, 'a situation which would make it impossible for an independent Scotland to run an independent economic policy.' Scotland has in fact three times as much non Scottish capital invested in her industry than even the mild Steuer report considered commensurate to maintaining some possibility of local economic control, finance capital on an international scale now playing the dominant role in integrating the international economy.[12] In drawing attention to how far Scottish finance capital is tied in a dependent way to the international capital market, John Scott and Michael Hughes, however, point to the allocative power of a small highly interconnected élite of sixty whose multiple industrial and financial directorships (and considerable political influence especially through the Scottish Council for Development of Industry) give them substantial control over economic decision-making in Scotland but whose freedom of action is constrained by the requirements of international capital as a whole. Thus, most of Scottish finance capital is invested outside Scotland, one-third of it in America. Scott and Hughes conclude that 'since the purely Scottish elements of the economy are dominated by internationalised capital which recognises no mere political or territorial lines drawn on the map, the success of a socialist strategy for Scotland depends upon its being able to counter the economic power of the upper class in Scotland and its links with international capital.'

John Foster traces the historical development of the dependency of the economy – and the far-reaching implications of the economic decline of the last fifty years. Today Scotland has neither the highly industrialised structure of Europe's industrialised regions nor the same concentration of services as the metropolitan areas. Agriculture has been run down to the extent that 42,000 agricultural workers are on the 95 per cent of the land which is non-industrial. Employment in the primary – agricultural and mining – sector was halved in the sixties and the manufacturing sector decreased by 6 per cent with losses most severe in textiles, shipbuilding, chemicals and metal manufacture. The assumption of the sixties was that a drastic alteration in the structure of our economy was required to fit in with market demands, through attracting highly capitalised technology-based industries to replace apparently declining industries, but today, as Bill Niven writes, drawing attention to the spread of recent closures, 'the problem of redundancy covers the whole spectrum of the economy, including the so-called growth sectors.' Neither Ravenscraig, electronics, the motor industry, nor Hunterston have provided the basis for a new Scottish economic revival. Indeed, it is ironic that the major growth points of the sixties were the relatively small manufacturing or service industries – food and drink, cars, clothing, insurance, banking and finance, tourism and local authority administration. Neither will oil provide a structural shift for Scottish economic progress, unless government policies are radically altered. Peter Smith draws attention to the close correlation between expected income from selling oil – £585m p.a. by 1980 – and Britain's rapidly escalating overseas debts and the projected Balance of Payments deficits – and he emphasises the consequent failure of government to stimulate British oil-related industries, citing the lost opportunities for British Steel and for the design, manufacture and installation of oil equipment generally – and the inadequacy of existing oil spin-off industries, particularly in petrochemicals. The result is a comparative dearth of available jobs, with only 30,000 workers – equivalent only to the average yearly loss in primary and manufacturing industries – likely to be directly employed in oil-related industries at the peak of expansion and only 250 more jobs projected for Clydeside. There will be little appreciable improvement in our standard of living[13] and indeed the pull of oil investment, requiring more capital yearly than all other manufacturing industry combined, has its own dangers – a shift of investment from other less profitable manufacturing to oil production. At the same time through their recent strike of capital the massive oil multinationals are, as one American finance journal put it recently, 'standing in the forefront of the fight against socialism in Britain.' In 1974 the Labour government decided not to nationalise oil: in 1975 they offered the oil companies tax concessions worth up to £1000m a year by the 1980s.

The trend of developments in Scotland is fairly clear. In suggesting that a dual economy between east and west Scotland was emerging, *The Economist* recently contended that Scottish industry must follow where growth beckons, while the West is forced to alter its distribution of industry. This would involve 'a planned movement of labour away from the West'.[14] A modified view of this picture was drawn by the Scottish Council (Development and Industry) in their report *A Future for Scotland*[15], proposing that the main axis of Scotland's development should be twisted from the old east-west Edinburgh-Glasgow line of their Oceanspan proposals to a new axis running from the north-east, from Aberdeen to Edinburgh, to the south west, between Glasgow and Ayr. But the net effect of such policies, even if there were to be a 2 per cent unemployment rate, is that one quarter of a million new jobs will have to be found by the mid eighties, in addition to jobs to replace the probable loss of another 100,000 by 1980. Existing manufacturing industry would continue to decline, while the service sector would increase by 100,000 jobs (and by $\frac{1}{4}$m. jobs by the year 2000). But apart from tourism and possible administrative decentralisation, the study emphasises that 'most growth in services cannot be seen as a generator of growth but as the result of growth generated in the basic manufacturing industries.'[16] What appears therefore to be likely is, as John Firn has shown for the sixties, a continued expansion of female and semi-skilled labour, to the detriment of existing skilled trades, and the creation of opportunities in industries and services whose growth prospects are dependent on the success of mainly the multinational in stimulating the Scottish economy as a whole. The service sector already has a concentration of low paid jobs, with 20 per cent of local male government workers, 40 per cent of retail trade workers and over 20 per cent of Government clerical workers paid at the supplementary benefits level. One group of economists has put it this way, that 'we should recognise that a solution to Scotland's economic difficulties will require significant geographical shifts in employment and production ... swimming with the tide is a lot easier than swimming against it.'[17]

Government policy since Toothill has rested on stimulating the growth of new, mainly science based industries through incentives and negative controls and on increased productivity in new and old. But both Frank Stephen and Bill Niven show that regional policy has in no way regenerated the Scottish economy. The performance of Scotland's economy has been best when the British economy as a whole has boomed. Only one third of the jobs required yearly in the sixties were actually created, at considerable public expense, in subsidies to private firms; development areas with labour surpluses have tended to attract capital intensive industry while labour intensive industry has moved to areas of labour shortage; three-quarters of the employment created has come from the establishment of new factories owned and operated from other regions or countries[18]; and while Scottish exports have outpaced the growth of British exports as a whole, expansion would appear to have been concentrated in areas of external control – one factor contributing with higher public expenditure to what has been calculated as a higher Scottish balance of payment deficit per head than in Britain.[19] Productivity agreements have themselves led to little expansion in jobs, one trade union leader recently telling a conference of industrialists that he 'did not know of any factory in Scotland where as a result of a productivity agreement a number of jobs were created.'[20] While public expenditure is nearly 50 per cent of the Scottish Gross Domestic Product, and while public investment has run at up to £750m yearly, the effectiveness of government intervention has been limited by their desire to maintain the existing market economy. Subsidies to private industry have formed 16 per cent of UK subsidies as a whole, while subsidies to public companies have formed 11 per cent of UK expenditure in that direction. There has been little use of state purchasing policy as an instrument of regional economic policy and little publicly sponsored and publicly controlled industry developed in Scotland, especially in the manufacturing sector, – despite the fact that a number of areas – the chemical industry, man-made fibres, electrical goods, pharmaceutical and plastic goods, aircraft, the industrial plant and mechanical handling sectors of engineering – have great potential for development in Scotland if publicly directed. The Scottish

Development Agency which was announced in January is 'to have the responsibility and resources to play a strong entrepreneurial role in identifying and promoting industrial modernisation, growth and development', but Stephen shows how much further the government should have gone in its proposals, particularly through a co-ordination of the presently anarchic allocation of public investment in Scotland[21] and the framing of an overall Scottish structure plan, in consultation with the trade unions, and with sensitivity towards the difficulties of Clydeside. Like the National Enterprise Board, the projected resources for the agency (£200m) will be insufficient.

Thus with American and British multinationals increasingly dominant in Scotland, and with the aims of public policy being to respond rather than lead, the Scottish economy is perhaps more subject to the influence of multinationals than any other similar industrial country. Consequently the economy is not only as Niven shows an unstable one, one of the first to suffer and the last to recover in times of depression, but also dependently subordinate to the international market, with an increasingly distorted and artificial division of labour, compounded by the massive export of capital. Firn highlights one aspect of this – the shortage of research and development work – but even *The Business Scotland* was moved to write, recently,

> 'Scotland in calling for jobs at any price for the past ten years has been engaged in turning its economy into one akin to that of a colony, i.e. one with a high level of external control, absentee-decision-makers and subsidiary technology.'[22]

Explanations tied to a vicious circle of low productivity, bad labour relations, low investment and poor entrepreneurship abound yet it is precisely the familiar tried formulas of wider incentives, tax reliefs, assured labour markets, growth areas and native entrepreneurs which are proposed. Thus public intervention will only be as effective as the efficiency of the private sector permits. The Scottish Council's policies for administrative decentralization and their research Institute's advocacy of devolved economic management (and reduced taxation) as a means of stimulating indigenous private industry[23] ignore the extent to which regional policy sprang from the failure of existing Scottish industry to compete and, as long as Scottish financiers maintain their dependent relationship to international capital, these policies may in the event not only increase the Scottish economy's powerlessness over world market trends but increase the relative underdevelopment of our economy. In a Scottish supplement the *Investors Chronicle* put it another way:

> 'If Scotland is able to achieve some degree of executive control over its own development, then what the Scottish Nationalists now call the "exploited province" could become the last frontier for the private developer.'[24]

For neither the promotion of smaller scale private industry, nor indeed the creation of a tax haven bears much relation to Scotland's own regional economic needs. Firstly, as Dr Stuart Holland has emphasised elsewhere, such policies might require massive, up to 75 per cent, subsidies in labour costs to remove the differential between Scottish and third world labour costs, or alternatively would require lower wages.[25] Secondly, these would lead to concentrating industrial expansion in growth areas, to the particular detriment of Clydeside. Scotland's economic difficulties are themselves regional. As the Scottish TUC have realistically pointed out, the loss of steel will by itself leave West Central Scotland, whose problems are as severe as any region of Europe, 'emasculated beyond all recognition.'[26] In his study of Glasgow, Cable shows that strategies must go further than the West Central Plan's proposed remit for a Strathclyde Economic Development Corporation, and involve publicly controlled industry. For the experience of the sixties shows that the market can no longer be seen as the efficient allocator of resources and indeed that the productive forces within our economy have outstripped the capacity of the market.

A Planned Economy

Niven argues that the public sector which employs 30 per cent of Scottish workers can be expanded in a manner which would dominate rather than respond to market requirements. Clearly the logic of present economic developments point in this direction. It is not just the demand of working people for a fuller share of the social product of their labour (and their collective power to resist the old formulas of unemployment and low wages in recession) – but also the cleft stick of labour displacing technology. The more automation there is, the greater is the need to deal with the social consequences by increased public expenditure; yet the more the government raises taxation, the more urgent is the need for more automation. Thus, increasingly, the private control of industry has become a hindrance to the further unfolding of the social forces of production. Consequently, Michael Barratt Brown has convincingly argued that increased state intervention in social and economic affairs implies that it is no longer realistic to envisage a socialist commodity exchange market in a transition from capitalism to socialism, and as a corollary, that an ever advancing technologically-based economy is not the only way forward for underdeveloped regions or countries.[27] Whether through investment in state owned industry in the central belt or through the application of intermediate technology, as Carter proposes, to the rural areas of Scotland, it is the erosion of the power of the market – and of the multinationals who now manipulate the market – to determine social priorities that is the forging ground for socialist progress.

The question of what socialist policies are required to meet the demands, skills and needs of Scottish working people raises the question of how the Scottish Labour Movement can force the pace of the advance towards socialism in Britain. Certain definite points of advance are obvious at a British level, although this does not rule out socialists pressing for an economic control co-extensive with economic devolution under a Scottish Assembly: the public control of industries essential to the provision of social needs and services, the priorities being building and construction, food and food processing, insurance and pensions; the industries essential to the planning services vital to the economy – the priorities being energy as a whole, land, banking and foreign trade; industries whose monopolistic position threatens the ability of society to plan its own future – the priorities being the taking over of the assets of the major British and American multinationals in Scotland; and industries essential to regional development – in Scotland's case shipbuilding and textiles being the obvious cases.

Smith details what is required for the planned control of energy – the nationalisation of all offshore oil and gas industry, the private sector of BP, the British sector of Shell and the Burmah-Castrol Group, to become part of a National Hydrocarbons Authority, and of GEC to form the basis of a national nuclear corporation. But he also shows that if the benefits from oil are to be such that long term economic growth is possible then ICI should be taken over to form a public chemical corporation. The proceeds from oil could themselves be transferred into a regional development fund. A second basic area is land, vital to the future of Scotland, in providing food, timber and other services. Jim Sillars suggests a concrete plan for taking land into public ownership, John McEwen and Ian Carter in particular show what a socialist land strategy could involve and how industry suitable to the skills and needs of the local population and available resources could allow substantial local control over Highland development. Ray Burnett suggests that one obvious step could be an elected Highlands and Islands Development Board. A third area, investigated by Scott and Hughes is the necessity for social control of the institutional investors who wield enormous financial power both in fostering privilege in our social security system and in controlling the economy. Two recent Labour Party pamphlets, 'Capital and Equality' and 'Banking and Insurance' propose how public control of banks, insurances and pensions companies, could have a two-sided effect: creating greater social justice in the social services and providing substantial resources for industrial investment.[28] Such a policy could be enacted without compensation and would in itself constitute a major erosion of the power of the British upper class. Public control to end the manipulative stranglehold of the monopolies would require a strategy to end the power of the British, American and European multinationals over the Scottish and British economies and in the event would require controls over foreign investment and trade, accepting a disengagement from a commitment to the free movement of capital in Europe. It would, as recent studies have indicated, require the forging of a new international economic framework based on long-term bilateral trading agreements for exchanges of goods and services and in the long run a payments union, possibly under the United Nations organisation, for clearing and extending such trade exchanges between nations, in particular, providing credit to underdeveloped countries.[29] Clearly, such a strategy is far more possible in Britain as a whole, given the substantial (and often underrated) industrial and financial assets of private companies. Britain has 140 of Europe's top 400 companies and the private sector has twice as much invested abroad as foreign companies and financiers hold in Britain.[30]

Workers' Power

But the demand for the economy to be directed according to people's needs requires that the need for meaningful work be prioritised. That involves a new and creative relationship between work, education and leisure, which breaks down the existing division of mental and manual labour and the extension of self-management at the work place. What has often been cited as an irresoluble clash in socialist theory between regulating material production according to human needs and the principle of eliminating the exploitative domination of man over man[31] can only be met through producers controlling the organisation of the production process. Thus it is precisely the surging forwards of demands by trade unionists for real control over the decisions affecting their livelihood that will be the point of departure for socialists. In his study of industrial democracy, Alex Ferry shows that the greater the influence workers have over their working lives locally, the greater will be the demand to reduce managerial prerogatives. Workers' control is impossible, he suggests, in a society which is not socialist – but controls developed in our present society which deny the logic of the market are the embryo of a future society. Clearly the proposals for workers shareholdings (which would, at 1 per cent in all of shares yearly, take at least fifty years to mean anything) and for worker directorships, are inadequate. But the most outspoken proponent of workers' control in Britain, Ken Coates, has seen the recent TUC proposals for industrial democracy – a supervisory board with 50 per cent workers representation having the final say in major investment decisions, closures, redeployment, location of plant and so on – as 'a cautious step in the right direction.'[32] The Labour MP for Motherwell has suggested that much more could be achieved if 'these representatives could by law be able to call for a ballot of employees as to whether they wish a scheme to be prepared for the conversion of the enterprise to workers' control.'[33]

Co-ordination is clearly required of workers' activity in different industries and unions. The trade unions themselves, as Ferry suggests, must take increased steps to link the demand for better conditions of work and pay with the pressure for increased control. If they do not, they will be left behind by rank and file action, such as we saw in Scotland in the last three months of 1974. Co-ordination, to be effective, must clearly be coherent around certain demands which allow a systematic advance in industry and society. The proposals of the Institute for Workers' Control at the time of the Upper Clyde Shipbuilders (UCS) occupations have more relevance than ever in today's recession.[34] A first condition set by trade unions in face of threatened unemployment would be that if the existing market is inadequate to support continuous production of ships, steel, textiles and so on, then the government should be bound to investigate the possibility of raising the whole level of international trade through state-guaranteed or negotiated trade exchanges. A second condition would be that if redundancies are inevitable, then government, locally or nationally, should organise alternative production to meet social needs – such as housing and community facilities – which private enterprise is failing to produce. An overriding condition would be that workers in such situations have the right to elect for control over their enterprises. Workers' control on an international scale is clearly an alternative to nationalism.

Political Options

In his recent investigation of the implications of increasing American control over the manufacturing industries of the regions of Europe, Poulantzas has suggested that the internationalisation of capital and the concentration of growth in the central axis of Europe will lead less to the much hoped for political integration of Europe but to the revival of nationalism in the peripheral regions. But the national financial and industrial élites who have increasing local political influence will remain so tied to international capital and to the international division of labour, he argues, that they are without the power to lead working people of the regions towards socialism. Scott and Hughes have suggested that such a description has relevance for Scotland, and Tom Nairn shows how the combination of uneven development and oil places Scotland in the vanguard of this revolt of the peripheral regions. While nineteenth century European nationalism was 'a response to enforced dilemmas of underdevelopment', new nationalism is 'a political response to the uneven development of capitalism which arises in areas on the fringe of metropolitan growth zones which suffer from relative deprivation, and are increasingly drawn to action against this'. But Scots people have never lacked a sense of nationhood or identity – and he shows how fatal it would be for socialists to dismiss nationalism as relevant only to early industrialisation or the escape from naked colonialism, to reject it 'as a disease which in Europe carried off millions, but in Scotland only affects the mind' in the hope that it will go away – or to suggest that recent Scottish nationalism is the peculiar affliction of a nation 'where the noblest of our ancestors died manfully for a dead cause – that of the ideas of the century before'.[35] A nationalism which has divided international communism – and now threatens to fragment modern Europe from the periphery inwards – is no mere bourgeois survival. But in asking what it means to Scots today – and in investigating a Scottish culture which is organised and lived, Nairn shows the manner in which Scottish social practices and expectations were moulded in an accommodated response to the development of the British market economy and Scotland's incorporation within the Greater Britain. The speed and timing of Scottish industrialisation in the nineteenth century, he demonstrates, precluded the characteristic European-wide response to modernisation in political nationalism. Rather the aspirations which nationalism satisfied elsewhere were met by the opportunities and success of the First Industrial Revolution and the British Empire and through a subculture which exhibited a marked dissociation from political action. Thus, Sir Walter Scott, tartanry, the Kailyard, militarism and the *Sunday Post* – all distinctive features of Scottish identity – were all part of one cultural impasse. This anomalous mode of economic and social development emphasises that there is self-defeating fantasy in a romantic nationalism[36] which poses the issue of Scots and non-Scots as value judgements. A successful British capitalism was the kernel out of which our ways of thinking and feeling were created. Accordingly, it has been the decline of British self confidence, particularly since 1945, that has created the conditions in which Scottish nationalism could become politicisable, and it is precisely our critical ability to 'rise above a weak national inheritance in a manner at once intellectual and universal in its aim' which could lead Scotland towards influencing a Europe 'which has crawled out of the abyss of the age of nationalism – or perhaps more precisely, has been dragged out of it forcibly by its own disasters and the combined powers of America and Russia.'

While drawing attention to the discontinuities of Scottish cultural development, both Tom Nairn and David Craig make us aware of the importance of examining the attempts of people to rise above the here and now which failed, the tendencies which were defeated and the aspirations which were crushed. Craig concentrates on the extent to which the best Scottish writers, 'who, seeing to the roots of the human condition, worked with whatever forces in society bid fair to win forwards to a better society', remain a radicalising and inspirational influence. MacDiarmid, whose love/hate relationship with the national geist has torn him between nationalism and communism, is properly described as existentialist: it is a vision of a fulfilled man in a fulfilled society, a Scotland purged of secondhandedness, the cultural equal of the rest of Europe. Like MacDiarmid, Gibbon saw it was imperative to transform the economic and social forces which bred deprivation and alienation in industrial and rural communities. But writing at the time that he did, he was unable to reconcile a socialist ideology which presumed ever-advancing industrialization with the continuance of a rural community incapable of mounting the organised and active resistance necessary. Only the industrial working class in the towns and cities were commensurate to the demands of history.[37] In studying the birth of the Scottish Labour Movement, James Young emphasises the variety of humanist, Marxist and libertarian strands in early socialist thinking, but pinpoints the importance which early socialists appreciated of forging links between socialism and industrial workers. This was a dialogue which succeeded – only to go sour in the hollowness of Labour's socialism in the late twenties. I argue elsewhere about the period which followed the First World War that where the issue of socialism against anti-socialism was fiercely contested – primarily in Scotland's central belt – there remains a definite commitment to social change which the Labour Movement must further harness, and that where, in the least industrialised areas of Scotland, the issue was little posed, and socialism made little headway, the challenge still remains for socialists not only to develop, as Carter and McEwen do, a social strategy which relates the use of land to the social conditions and needs of these areas, but also to develop an organisation and momentum.[38] We would be wrong to underestimate the experience and the education which has led particularly the industrial workers of Scotland to reject implicitly if not explicitly the values of a capitalist society.

Socialism and Nationalism

Nairn's prediction that nationalism may now be unstoppable arises from greater pessimism – the failures of socialism and the successful adaptation of international capitalism. But the question of commitment is, of course, a matter of agency and choice – where men and women can and will make their own decisions in a situation where neutrality is increasingly impossible. In terms of imperatives, the debate about attitudes to nationalism and the SNP in particular is made possible by the inclusion of articles from Burnett, Tait, McGrath and Edwards. Bob Tait argues that a breakaway Scotland would be one important lever against multinational capitalism, and that there is a radical base within the SNP which would be enhanced through its fusion with socialists. Owen Dudley Edwards sees possibilities in Scotland today for a more open society – multicultural, democratic and less corrupt – where a freedom which is permanently extended can allow an avoidance of the tragedies of events in Ireland. But both Ray Burnett and John McGrath suggest that the policies of the SNP are tied to the existing consensus, that their rise to power in an independent Scotland would lead to a Scottish corporate state little different in substance from the present British state and that a socialist strategy must begin with a direct and conscious attack on international capital. The SNP, they conclude, is no place for a socialist to be.

Clearly, from what I have already said, my own view is that the SNP's 'new politics' which 'reject class warfare' presumes the familiar priorities of wealth and power over people. Their incoherence over the impact of multinationals on the Scottish economy, their rejection of the public ownership of land, oil and basic industries and their corresponding faith in incentives, and local entrepreneurship is a familiar blend of the old well worn formulas, which assumes the subservience of Scottish workers to private international controls.[39] Their programme for a redistributive 'Scottish social justice' – which includes an inadequate £25 minimum wage, 4 per cent mortgages, a rates holiday, and strengthened occupational and pensions schemes – not only wrongly assumes that economic growth within a mixed economy will satisfy the divergent claims of all classes for a share in the growth but, as one study pointed out, will not substantially extend social services provision.[40]

The SNP's divorce from the Renaissance Movement, reflecting what Nairn terms the Scots 'cultural schizophrenia', has led MacDiarmid to criticise the party for having 'no concern with things of fundamental importance, with the great spiritual issues underlying the mere statistics of trade and industry', and corresponds with the findings of one study of local SNP officials which recorded that 'feelings of Scottish identity do not exist within the context of a flourishing Scottish culture'[41]. This is, of course, partly a reflection of the SNP's leadership which has brought the professional and commercial middle class back into politics *en bloc* for the first time since the Liberal decline; there were only two working class SNP candidates at the October election while there were ten managers.[42] Recent studies of the existing membership of the SNP emphasise that their support comes mainly from those groups least likely to be influenced by the collective experiences and loyalties of the workplace and working class communities.[43] This does not mean that disaffection with Labour will not lead to what one writer has called 'avarice in the absence of a clear and effective solution to long standing economic difficulties'[44]. But the challenge is how at one and the same time the Scottish Labour Movement can reach to the roots of people's experiences and aspirations and lead the demand for change.

The Way Forward

There are as many Scottish roads to socialism as there are predictions of Britain's economic doom – but most of them demand three things: a coherent plan for an extension of democracy and control in society and industry which sees every reform as a means to creating a socialist society; a harnessing of the forces for industrial and community self-management within a political movement; and a massive programme of education by the Labour Movement as a whole. Gramsci's relevance to Scotland today is in his emphasis that in a society which is both mature and complex, where the total social and economic processes are geared to maintaining the production of goods and services (and the reproduction of the conditions of production), then the transition to socialism must be made by the majority of people themselves and a socialist society must be created within the womb of existing society and prefigured in the movements for democracy at the grass roots.[45] Socialists must neither place their faith in an Armageddon of capitalist collapse nor in nationalisation alone. For if the Jacobin notion of a vanguard making revolution on behalf of working people relates to a backward society (and prefigures an authoritarian and bureaucratic state), then the complexity of modern society requires a far reaching movement of people and ideas, acting as a stimulus for people to see beyond the immediacy and fragmentation of their existing conditions and as a co-ordinator for the assertion of social priorities by people at a community level and control by producers at an industrial level. In such a way, political power will become a synthesis of – not a substitute for – community and industrial life. This requires from the Labour Movement in Scotland today a positive commitment to creating a socialist society, a coherent strategy with rhythm and modality to each reform to cancel the logic of capitalism and a programme of immediate aims which leads out of one social order into another. Such a social reorganization a phased extension of public control under workers' self-management and the prioritising of social needs set by the communities themselves – if sustained and enlarged, would, in E. P. Thompson's words, lead to 'a crisis not of despair and disintegration, but a crisis in which the necessity for a peaceful revolutionary transition to an alternative socialist logic became daily more evident.'[46]

But the dynamic must come from the existing layer of thousands of committed socialists in Scotland today, firstly through a most obviously democratic and accessible Labour Movement co-ordinating its work with the trade unions (beginning with factory branches) and with street committees, and secondly, through a concerted programme of political education. The early Scottish socialists believed that the bridge between their utopian ideals and the practical politics under which people suffered must be built in a massive programme of education and propaganda. Today in Scotland we have no daily or weekly specifically Scottish political newspapers, no socialist book club, no socialist labour college, no workers' university, and only a handful of socialist magazines and pamphlets.[47] We need all of these now.

It is only within such a reinvigorated socialist strategy that we can appreciate the possibilities if existing and proposed structures of government. Devolution has been all things to all people – the halfway house between Westminster rule and a Scottish independence that will take us from rags to riches; the insertion of a sixth tier of government which threatens to make us the most over-governed country in Europe; and a fundamental extension of democracy whose every detail is of prime concern and importance. David Gow shows that devolution was intended as none of these. In proposing 'the grafting on' of a devolved Assembly to 'a constitution which in its essentials has served us well for some hundred years', Kilbrandon's main concern was to improve checks and balances against the centralisation of governmental powers, so to secure 'a restoration of public confidence' in government. Kilbrandon neither investigated the influence of private power over the political system (and how the state had assumed greater social and economic powers as a response to the failings of a market economy) nor did he examine how 'the demand for more control over our affairs' might lead to greater participation.[48] Sitting at the time of UCS and heightened industrial militancy in Scotland, his limited research made him unaware, as one MP has put it, that 'much of the feeling in Scotland was closer to the demands for workers' control than to the classic nationalism of the nineteenth century'[49]. At the best, Kilbrandon assumed that the continual interaction and competition of pressure groups in the political area will sustain a progressive enlargement of the possibilities of their goals being fulfilled; at the worst, all will be compromised and none will be satisfied, with such diffuse groups as the unemployed, the elderly, the low paid with little power to wield, losing out. Consequently, more than 40 per cent of those questioned in a recent ORC poll believed an Assembly in Edinburgh would be as unreachable as government in Westminster, and did not regard it as a priority.

The question is not how men and women can be fitted to the needs of the system – but how the system can be fitted to the needs of men and women. Labour must respond to 'the demand for more control over our affairs' not by asking how 'minimalist' or 'maximalist' Assembly powers can avoid separation nor by becoming masters of the last ditch – resisting change until it becomes inevitable – but by deploying every available level of government to increase the control working people have over their lives. For while politics should never be reduced to 'Assemblyism', they will not readily be separated from it. Ronald Young shows that it is how socialists approach their responsibilities as elected representatives that is the crucial question, local government reorganisation offering the opportunity for a more integrated and politicised approach to meeting needs generated by local communities and a more variable framework for relating local to national issues. The Assembly, too, can not only enable regions and districts to make radical innovations but also be a focus for and co-ordinating point for formulating Scotland's priorities. It offers a political control over the Scottish Office, leading both to an accountability and to a freer flow of information. It allows the framing of distinctly Scottish policies to meet social needs and requirements. It gives Scottish people a focus for bargaining with Westminster and Brussels – and gives Scottish socialists the chance to lead and influence other regions and other countries. Finally, it is clear that there is nothing inherently anti-socialist about economic devolution as long as Scottish Labour insists on genuine economic control in devolved areas nor is there anything remiss about taxation powers in relation to democratically decided levels of public service provision. But the real opportunity which present events offer is, of course, something more. It is the challenge to force the pace towards socialism in Britain as a whole, and to reinvigorate the labour movement in Scotland from the workplace and community outwards. Scotland's socialist pioneers, Hardie, Smillie, Maxton, Maclean, Gallacher, Wheatley and others, knew that socialism would not be won until people were convinced of the necessity for social control. The Scottish Labour Movement is uniquely placed today to convert the present discontent into a demand for socialism: we will fail only if we ignore the challenge.

References
1. The Red Paper on Scotland is the second Red Paper, intended as a forum for the left to express their views on immediate issues. The first was **The Red Paper on Education**, (ed. Bob Cuddihy), Edinburgh 1970. References in this introduction to articles or writers which are not footnoted relate to contributions to this collection.
2. **Sample Census of Great Britain, 1966**, Economic Activity Part 3 (HMSO, 1969), Table 30, pp. 421-422.
3. **The Economist**, September 29th, 1973, Scottish Supplement p.5. For a discussion of wealth inequalities in Britain, see A. Atkinson, **Wealth and Inequality in Britain** (1973). For Scotland, see also L. Wright in J Wolfe (ed), **Government and Nationalism in Scotland** (1969), pp. 140-152.
4. Board of Inland Revenue, **Inland Revenue Statistics (1973)**, p. 105 (HMSP, 1973).
5. Board of Inland Revenue, **Survey of Personal Incomes 1970-1971** (HMSO, 1973), pp. 102-3.
6. Scottish Development Department, **Bowhouse, Alloa.** Bowhouse is an area comprising 4 per cent of the population of Clackmannan County. The study was reported in the **Guardian** in October, 1974.
7. Public Expenditure in Scotland in **Investing in Scotland's Future,** p.177 (Scottish Council (Development and Industry), 1974).
8. Educational Priority: **A Scottish Study** (HMSO, 1974) esp. p.184-5.
9. See in particular R. Furbey in **Social and Economic Administration** Vol. 8, No. 3 (1974) pp. 192-221.
10. **Public Expenditure in Scotland, op.cit.,** pp. 160-164. I am grateful to Andrew McPherson for information on university entrants.
11. This was proposed at the time of the Upper Clyde Shipbuilders (UCS) work-in by M. Barratt Brown in **UCS: The Social Audit** (Institute for Workers' Control, Pamphlet No. 26, 1971).
12. M. Stuer, **The Impact of Foreign Investment**, cited in Firn.
13. See, for example, the **National Institute Economic Review** (NIESR, Aug. 1974).

14. **The Economist**, op.cit., esp. pp. 27-35 and p. 47
15. **A Future for Scotland**, Report by the Scottish Council (Development and Industry) (1973).
16. **ibid.,** p. 109.
17. D. McKay and A. McKay in **The Scotsman**, February 5th, 1975, p.10.
18. The most recent study is Moore and Rhodes, Regional Policy and the Scottish Economy in **Scottish Journal of Political Economy**, Vol. XXI, No. 3 (Nov. '74), p. 233.
19. SCRI, **Scottish Manufactured Exports** (Nov. 1974), Begg et al 'Scotland's Balance of Trade' (1975), pointing to substantial trade deficits, quoted in **The Scotsman**, February 6, 1975.
20. Ray MacDonald in **Scotland's Goals** (1974), p. 54.
21. Alexander has recently shown that only 15 per cent of total public investment is 'decided primarily by Government itself', **Investing in Scotland's Future**, p. 58. See also the more radical proposals of the Labour Party (Scottish Council) in **Scotland and the N.E.B.** (1973). The STUC have rightly not accepted that 'the role of the SDA should be to help private enterprise to help itself.' Instead they urge that the SDA be given compulsory purchase powers to take equity in Scottish companies, and have one nationalised bank under it. **Scotsman**, February 28th, 1975.
22. **Business Scotland**, February, 1975, p.48.
23. Scottish Council, **Memorandum** submitted to Lord Crowther Hunt (1974). It is discussed by David Gow. For a variation of the theme, see **The Guardian,** Devolution Supplement (1975) and **The Financial Times** Scottish Supplement, Nov. 11th, 1974. Scottish Council Research Institute, **Economic Development and Devolution** (1974), which favours wide ranging economic powers under an Assembly. See also, **The Scotsman**, Annual Financial Review, May 21st, 1974, 'Need for Scots to invest in Scotland.'
24. Scottish Finance and Investment, p.15, Supplement to **Investors Chronicle**, October 25th, 1974.
25. Evidence to House of Commons **Expenditure Committee**, Vol. 42, No. 18.
26. **Business Scotland**, February, 1975, p.45.
27. M. Barratt Brown, **From Labourism to Socialism** (1972); see also P. Mattick, **Marx and Keynes: The Limits of the Mixed Economy** (1969).
28. Labour Party, **Capital and Equality** (1972) **Banking and Insurance** (1973).
29. M. Barratt Brown, **Europe: Time to Leave and How to Do It.** (1974). See also Labour Research Department, **The Menace of the Multinationals** (1974), p. 310.
30. B. Rowthorn, Britain and the World Economy: Breaking the Chains in **Marxism Today** (Aug. 1974), p. 231-2. See also **Tribune**, January 24th

1975.
31. L. Kolakowski in **The Socialist Idea: A Reappraisal** (1974) pp. 18-35. For a reply, see E. Thompson, in **Socialist Register, 1973**, pp. 1-100.
32. **Tribune**, 31st January, 1975. p. 3.
33. Dr. J. Bray, **Towards a Workers Managed Economy** (Fabian Tract 430, 1974), also G. Radice **Working Power** (Fabian Tract 431, 1974) and K. Coates, **Essays on Industrial Democracy** (1971) and others.
34. M. Barratt Brown, op.cit., see also R. Murray, **UCS: The Anatomy of Bankruptcy**(1972).
35. P. Geddes, Quoted in H. Hanham, **Scottish Nationalism**(1969). Also N. Poulantzas, The Internationalisation of Capital and the Nation State in **Economy and Society, Vol. 3, No. 1, 1974 pp.145-179.**
36. For an extreme view of Tom Nairn's observations, see A. Jackson The Knitted Claymore in **Lines Review**, No. 3 (1971), 'For what could be done to restore Scots as a full living language ... expel the (million) English ... all papers and media in Scots ... all education in Scots ... the colloquial/literary split has more of a class basis than the national one' (p. 21-23).
37. See in particular L. Gibbon on Land and Glasgow in **A Scots Hairst** (ed. Monro), (1967), and H. MacDiarmid in **Whither Scotland** (ed. Glen) p. 239-240.
38. G. Brown, **Socialism and Scottish Political Change**, unpublished paper (1974) see also M. Dyer on Highland voting behaviour in **The Scotsman**, November 1974, where he describes the substantial working class vote as 'the silent voice of Scottish politics.'
39. SNP's Four Point Economic Plan, September 1974, which acknowledges continued support for multinationals in Scotland. See also SNP Manifestos, Feb. and Oct. 1974, and articles by D. Crawford in **Scots Independent**.
40. I. Levitt, **The War on Poverty**, unpublished paper, (September, 1974).
41. H. MacDiarmid, **A Political Speech** (1972) p. 9 and J. Schwartz in **World Politics** (1970) p. 496.
42. Labour Research Department Fact Service, 5th Oct. 1974. Of sixty-one candidates for which background information was found, 20 were teachers and lecturers, 10 were managerial staff (inc. one 'industrial therapy manager'), 8 were solicitors, 5 were professional engineers, 4 were journalists, 4 were professional workers, 5 were company directors, and 3 were farmers. There was one joiner and one Post Office engineer.
43. Drieux, Univ. of Sorbonne, Thesis on S.N.P. (1974). See also I. McLean and J. Blochel and D. Denver in **Political Studies** (1970) and J. Cornford and J. Brand in Wolfe, **op.cit.**
44. McKay and McKay, **The Scotsman**, 7th February, 1975.
45. Gramsci, **Prison Notebooks** (1971); see also L. Magri, Problems of the

Marxist Theory of the Revolutionary Party in **New Left Review**, 1970, pp. 97-128 and also P. Friere, **Pedagogy of the Oppressed** (1972) and **Cultural Action for Freedom** (1972).
46. E. Thompson in **Socialist Register 1973** p. 52 which is an updating of his brilliant articles, Agency and Choice in **New Reasoner**, 1958.
47. The various left parties recent pamphlets on Scotland's problems are: Labour Party (Scottish Council), **Scottish Manifesto** (1974) and **Planning for Prosperity** (1975); the Communist Party, **For a Scottish Parliament** (1974) and the periodical **Scottish Marxist**; International Socialists, **Socialism or Nationalism** (1974); International Marxist Group, **Scotland, Labour and Workers Power** (1974); and **Scottish Vanguard**, periodical of the Workers Party of Scotland.
48. Royal Commission on the Constitution (1973) esp. paras 311 and 396. J. P. Mackintosh summed up Kilbrandon well, 'The Commissioners did not search for alternative remedies in their questions; they asked almost exclusively about devolution', **Political Quarterly**, (Jan. 1974) p. 117. See also J. Stanyer in **Social and Economic Administration** (1974) who concluded that 'The Commissioners ... are not stemming the tide, but swimming even faster with it.' P.147.
49. Norman Buchan in **Whither Scotland** (1971) p. 89.

Also available from Socialist Renewal

New Labour's Attack on Public Services - *by Dexter Whitfield*

New Labour is creating markets in public services on an unprecedented scale. Action by alliances of trade unions, community organisations and civil society organisations is urgently required to protect education, health and other vital public provision.

Price: £11.99 | ISBN: 0 85124 725 3

Nuclear Reactors: Do we need more? - *by Christopher Gifford*

The case for 'fast-tracking' new nuclear reactors has not been made, argues an experienced health and safety inspector who has probed nuclear's safety record.

Price: £2.00 | ISBN: 0 85124 154 9

The Social Europe We Need
by Robin Blackburn, André Brie MEP, Ken Coates, Christina Beatty & Stephen Fothergill

The European Union is, at present, the only global entity with an economic weight & political potential equal to that of the United States. We should ensure that Europe represents a different social model to that of the United States, instead of becoming more like it.

Price: £9.99 | ISBN: 0 85124 703 2

Railtracks in the Sky: Air Transport Deregulation and the Competitive Market - *by Peter Reed*

An International *managed* regime for air transport could be made to work, starting at the European Union level.

Price: £9.99 | ISBN: 0 85124 671 0

Physician Heal Thyself - The NHS needs a voice of its own
by Duncan Smith

An NHS Staff College would allow directions from the top to be complemented by feedback from below. The NHS' former Chief Training Officer outlines the case.

Price: £3.00 | ISBN: 0 85124 667 2

The Captive Local State: Local Democracy under Seige
by Peter Latham

What is happening to local councils? The author, Secretary of the Labour Campaign for Open Local Government, analyses the changes under New Labour.

Price: £2.00 | ISBN: 0 85124 651 6

Safety First? Did the Health & Safety Commission do its job?
by Christopher Gifford

"This pamphlet is about how opportunities to protect people from abuse were taken or missed in the 25 year history of the Health & Safety Commission."

Price: £2.00 | ISBN: 0 85124 652 4

Spokesman Books, Russell House, Bulwell Lane, Nottingham, NG6 0BT
Email: elfeuro@compuserve.com | Tel: 0115 970 8381 | Fax: 0115 942 0433

www.spokesmanbooks.com